COLLINS AURA GARDEN

D0268703

# CLEMATIS

ETHNE REUSS CLARKE

HarperCollins*Publishers*

**Products mentioned in this book**

| | | |
|---|---|---|
| Benlate* + 'Activex' | contains | benomyl |
| ICI Slug Pellets | contains | metaldehyde |
| 'Sybol' | contains | pirimiphos-methyl |

Products marked thus '*Sybol*' are trade marks of Imperial Chemical Industries plc
*Benlate** is a registered trade mark of Du Pont's
**Read the label before you buy: use pesticides safely.**

**Editor** Emma Johnson
**Designers** James Marks, Steve Wilson
**Picture research** Moira McIlroy

This edition first published in 1988 by
Harper Collins Publishers
London

Reprinted 1989, 1990, 1991

© Marshall Cavendish Limited 1985, 1988

All rights reserved. No part of this publication may be reproduced or transmitted in any form or by any means, electronic or mechanical, including photocopying, recording or any information storage and retrieval system now known or to be invented without permission in writing from the publisher and copyright holder.

**British Library Cataloguing in Publication Data**

Clark, Ethne Reuss
Clematis.——(Collins Aura garden handbook).
1. Clematis
I. Title
635.9'33111 SB413.C6

ISBN 0-00-412373-5

Photoset by Bookworm Typesetting
Printed and bound in Hong Kong by Dai Nippon Printing Company

*Front cover: Clematis* 'Etoile de Malicorne'
*Back cover: C.* 'Ville de Lyon' and 'Perle d'Azur'
Both by the Harry Smith Horticultural Photographic Collection

# CONTENTS

# INTRODUCTION

The genus clematis contains at least 300 species distributed throughout the temperate zones of the world. Most are deciduous, the exception being the tender *armandii*, and *cirrhosa balearica*, a native of the Balearic Islands; these have attractive evergreen foliage and dainty flowers which make an attractive feature in the winter garden.

Although most of the species are of little value in a domestic garden, some are welcome for their scent, such as the *montanas* and *flammula*, while others, like *orientalis* and *tangutica*, are valued for their attractive silky seedheads.

The majority of the species have quite small flowers, but they are very fast growers, and so are especially useful for covering sheds and garages and for providing quick-growing screens in the garden. Their greatest contribution, however, lies in the wide variety of hybrids created over the last 100 years. By crossing species, hybridists have created clematis that will allow us to have flowers from spring to early autumn, in beautiful shades from deepest purple through rose to white. Some have single flowers opening flat to as much as 20cm (8in), and others are double and look like feathery pompons.

Clematis do not have true petals; the flowers are made up of enlarged sepals. Blooms last from three to four days each, generally opening a shade darker than the true colour, then fading gradually as the flower dies. On a mature plant in full flower this gradation of colour is an attractive feature in itself. Sometimes, if there is a cold spell in early spring, the sepals may open green and never attain their true colour. Therefore the early flowering hybrids should not be planted in too shady a spot; a daily dose of warming sunlight should prevent the unwanted greening. On the other hand, pale flowered hybrids that open later in the season should be

During the sixteenth century, many plants were introduced to Britain from other parts of the world, including several species of clematis. It is thought that *C. cirrhosa balearica* (above) came from the Balearic Islands in 1590. It was not until the nineteenth century that large-flowered hybrids, such as 'The President' (left) were introduced.

given north-facing sites to prevent the sun fading their delicate tints.

Clematis are members of the *Ranunculaceae* family, and are relatives of the peony, delphinium and buttercup. Like these plants, they will grow best where they have a moist, cool root run.

The hybrids are divided into groups according to their time of flowering: the Florida and Patens groups flower in May and June and then have a small show again in September. *C.florida* is a native of China, where it was discovered by Dr Augustine Henry nearly 100 years ago. The Lanuginosa group is derived from the type plant *C.lanuginosa*, also a native of China, introduced to the West c.1850 by the great plant collector Robert Fortune. Hybrids in this group begin flowering in June, ending in September or October. They all produce flowers on growth made during the previous season.

The Jackmanii and Viticella groups flower continuously throughout the months from June to September and October on the current season's growth. *C.jackmanii* was the first hybrid clematis introduced, raised almost 125 years ago by the nurserymen Jackman and Sons of Woking. But *C.viticella* has a longer history; it was included by John Gerard in his *Herball*, published in 1633, in which he gives its common name as 'Ladies Bower'.

The herbaceous species of clematis are a delightful addition to the flower border, and range in height from 45cm-1m (1½-3ft). Most have sweetly scented flowers. The taller-growing types should be staked with twiggy branches, or they can be treated as climbing clematis.

It is according to these divisions that it is determined how to prune the plants to produce the strongest growth and most flowers in a season. This is discussed in the chapter on pruning.

**Buying plants** A good plant is not necessarily the largest or the tallest offered for sale in a garden centre or nursery. The main thing to look for is healthy growth, and this cannot always be judged from the appearance of the foliage, especially if you are buying a plant when the foliage is dying back in the autumn. The months of October and November are the best times to plant clematis. At this time of year look for plants that have sturdy main stems. Otherwise the clematis may be planted in early spring before new growth starts, in which case look for fat, well-formed buds in the leaf axils as an indication of future healthy growth.

Before you decide which clematis to buy, think about how you plan to use it. For example, it's no use selecting one that has to be pruned to the ground each spring if you want a permanent screen or cover – for that you should choose one of the evergreen species. Conversely, if you want to grow a clematis as a feature in the flower border, either up a cane support or through another plant, don't choose a rampant grower that would smother all the surrounding plants in one season.

Clematis is certainly the 'Queen of the Climbers'; it is difficult to think of any other plant which gives such magnificent flowers and (with a little judicious selection) over such a long season. It is well worth the time and small expense involved to send for the catalogues of specialist clematis nurseries. They have the widest selection available, and the accompanying descriptions will be an accurate and invaluable guide when making a choice.

# GROWING CLEMATIS

The ideal time to plant any clematis is in the autumn. The plant is encouraged by the warm soil to put its energy into making new roots, and it will have settled in ready to begin new top growth in the spring. Also, the soil tends to be more moist in the autumn so the risk of damage through lack of water is lessened.

Limy soils are ideal for clematis, but they will grow in any good garden soil. Once you have chosen the planting site, it is a good idea to take time to prepare the soil. Clematis like deep, well-dug, moist soil, so consider the quality of soil in your garden. If it is heavy clay you will have to add peat; loose, sandy soils will be improved by the addition of loam or organic matter such as 'Forest Bark' Ground and Composted Bark. Check the new plant in the pot; if the root ball is dry put it to soak in a bucket of water for at least an hour. If the plant has been container-grown in a soil-less mixture of peat and sand, you should leave it to soak overnight to be certain that the roots are well moistened before planting.

Clematis should be placed at least 40cm (15in) from walls and 60cm (2ft) from the base of a tree. This is because a wall will rob the clematis roots of necessary moisture, and trees take much of the available nourishment from the surrounding area. Dig a hole at least 45cm (1½ft) in diameter and to a depth of the same. Break up the bottom of the hole and fork in two buckets of well-rotted manure or compost, to which has been added a handful or two of 'Forest Bark' Ground and Composted Bark. Fork over the sides of the hole, then replace some of the soil, covering the manure or compost with at least 10cm (4in). It is important to prevent new roots coming into direct contact with the manure or compost. Mix a generous amount of peat into the returned soil.

Hybrid clematis have a much sturdier root system than the species plants; the difference is similar to that between baling twine and fine thread. So when removing the pot before planting, you must be careful not to damage the root ball of a species plant. Gently loosen or unravel the roots of a hybrid before planting.

Place the plant in the hole so that the crown of the root ball is at least 10cm (4in) lower than the top of the hole. This will encourage new roots and shoots to form from below the soil level. Be very careful of the main stem when planting, and leave it attached to the supporting cane with which it came; incline the plant towards the support and leave the cane in place until the plant has become established and is growing into the support. Now replace the soil, adding several good handfuls of 'Forest Bark' Ground and Composted Bark. Tread the soil down around the plant to firm it, then water well.

Clematis grow best with their roots in the shade and their heads in the sun. This may sound an impossible situation to provide, but all you need to do is cover the base of the plant with a layer of peat or compost mulch. Alternatively, you can prop up a few tiles against each other, surrounding the base of the stem. But probably the best solution is to plant a small evergreen shrub near the base of the clematis; choose one that will not grow more vigorously

1. Dig a hole; loosen the sub-soil to ensure free drainage. Put in well-rotted manure or compost.

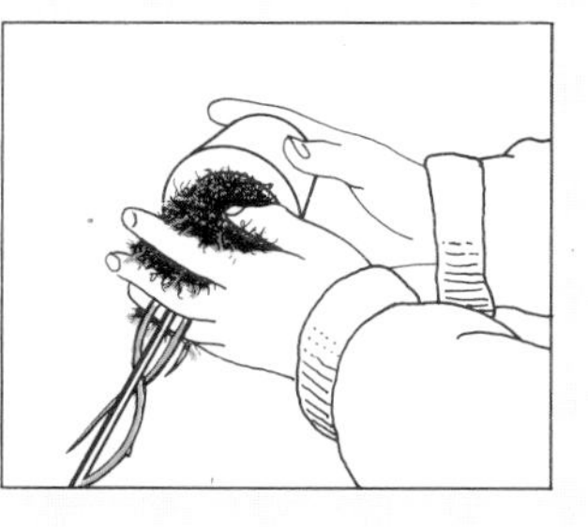

2. When removing the plant, support the main stem with your fingers.

3. Gently loosen the roots of hybrid plants; do not disturb those of species.

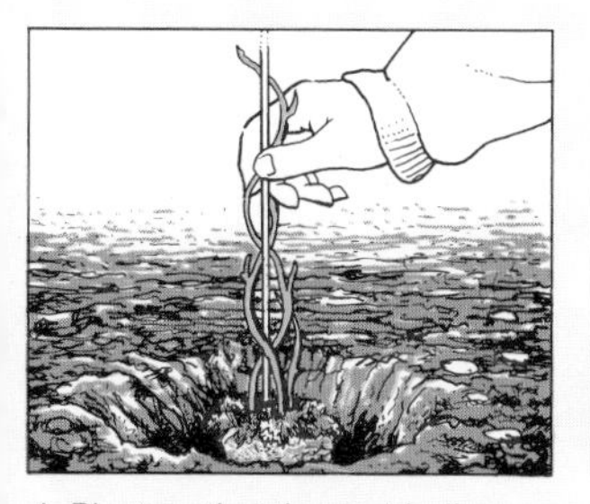

4. Plant so that the root ball crown is 10cm (4in) below soil surface.

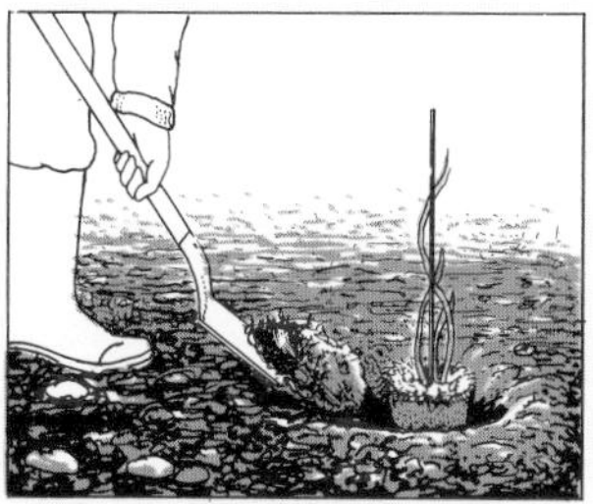

5. Return the soil, adding peat and some bonemeal.

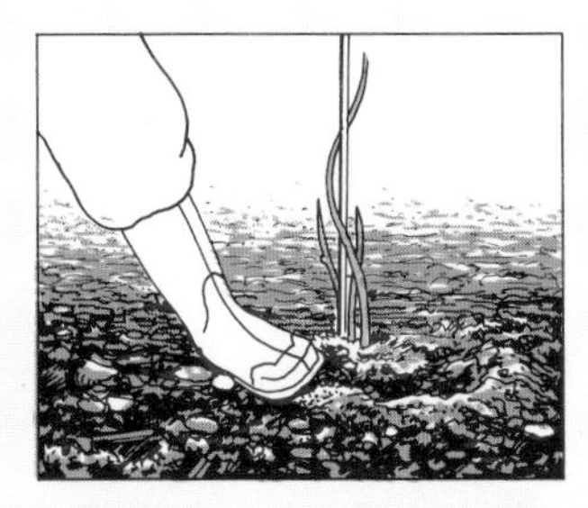

6. Firm the surrounding soil with your heel.

7. Water generously until well-established and apply a general-purpose liquid fertiliser.

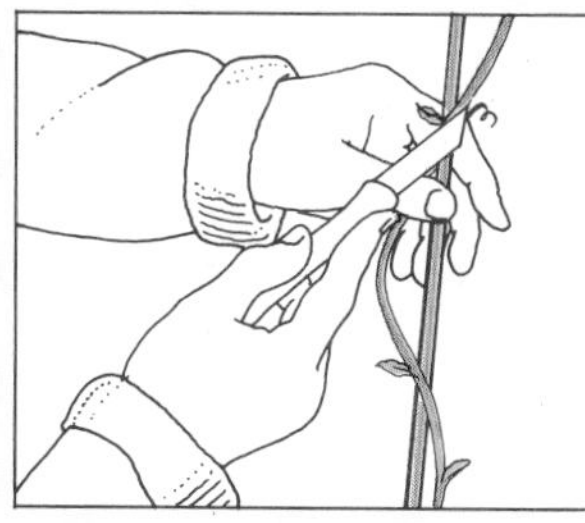

8. The following spring, prune hybrids back to a pair of healthy buds.

9. If planting in the sun, keep the roots cool and moist with stones, tiles or mulch.

than the clematis or rob it of food.

It will be a month or more before the plant becomes established in its new home and its root system is capable of drawing moisture from the soil. Help it along by watering frequently and thoroughly. A large flowerpot sunk into the ground 15cm (6in) from the base of the plant is a useful aid; fill it with water, which will drip steadily through the drainage hole directly to the roots. Liquid feeds of ICI Liquid Growmore can also be given using this method.

If you suspect there is a chance that the young plant could be damaged, either by careless gardening practice or by the unwanted attentions of the family pet, it can be protected by wrapping a piece of wire mesh or heavy corrugated cardboard around the base of the plant. Should the plant be broken off near the base, the buds on the stem below the soil level will be stimulated into growth and, while it will take longer for the clematis to achieve the desired size, the plant is not lost.

**Supports** Unlike climbing plants which attach themselves to the support by means of suckers, such as ivy, or by tendrils, like the grapevine, clematis climbs by twining its leaf stems around the support. Therefore, to have a clematis grow

TOP Fix trellis at least 2cm (1in) from the wall.
ABOVE Wooden grid trellis, supporting 'Royal Velours'.
RIGHT A hybrid, growing through other plants can look attractive.
FAR RIGHT 'Mrs Cholmondeley' shown to good effect on a trellis.

up a wall or similar flat surface you must fix wires or netting to provide the necessary framework around which the leaves can twist. Use either vine eyes or masonry nails, hammered into the brickwork and threaded with plastic-coated garden wire to make a grid of squares 25 × 25cm (10 × 10in). Ready-made wire or plastic-covered trellis is also available from garden centres and is a suitable (but slightly more expensive) alternative. One advantage of ready-made trellis is that it can be detached from the wall easily to allow for routine painting or other maintenance. Whichever method you choose, the wire support must be at least 2.5cm (1in) from the wall to allow space for the clematis to twine in and out as it grows up.

Natural supports are particularly attractive: the association of clematis and another wall-grown plant can be most effective, and flower display can be prolonged if, for example, you train a late-flowering hybrid such as 'Marie Boisselot' through an early flowering climbing rose or wisteria.

Clematis grown through trees or shrubs make a beautiful display. The *montana* clematis are ideal for growing through large trees; they do not need pruning and will quickly grow up to 18m (60ft) – a shower of starry blossoms dripping through early spring foliage. Some of the large-flowered hybrids are excellent for growing through shrubs. Dwarf conifers, helianthemum, lavender or any of the grey-leaved shrubs are suitable candidates for this treatment. Because the shrubs are low growing, pruning the clematis poses no problem. *Jackmanii* and *viticella* clematis that flower during summer can be trained through spring-flowering shrubs, and late-flowering ground covers like erica or calluna can be given a mantle of flowers by growing clematis hybrids through their spikey branches. More detailed information on how to plant clematis in these special situations will be found in a later section.

**Early training** Most of the species clematis used in gardens can be left to grow on until they fill the space. The only assistance they require in the first few years is making sure that the stems can easily reach the support and that new growth is distributed evenly over the framework provided. This is important as some of these plants can make heavy growth in their second year which, without supervision, twines round itself until the whole plant becomes a tangled mass.

All of the large-flowered hybrids should be cut back hard in the first spring after planting, cutting the main stem back to the bottom pair of buds. This helps the plant to make a good fan of stems that will carry the flowers in an attractive open display rather than in a tight jumbled mass at the top of one long stem.

# PRUNING

To get the most from your clematis it is important to know to which group it belongs and to understand its pruning requirements; pruning directly affects the flowering. Specialists divide clematis into three groups.

The first group is the species clematis, including the *montana*, *macropetala*, *alpina*, and evergreen clematis and their cultivated forms. These do not have to be pruned regularly except to keep them neat, and on older plants to rejuvenate them. Generally these produce their flowers before the end of May, the flowers appearing in clusters at the end of short stems which sprout directly from the leaf axils.

Any pruning should be done immediately after flowering, and never later than the end of June. Cut out any weak or dead stems and remove unwanted branches if the plant has outgrown its space. This will encourage new growth during late summer, which will ripen and produce flowers the following spring. After pruning, be sure that the remaining stems are securely tied into the supports.

The second group is the early flowering hybrids, which produce their flowers on old wood. This is growth made and ripened during the previous season. They begin flowering before the end of June, a single bloom to each stem; there is a second show after the first flush of flowers, but the blooms are smaller. These clematis should be pruned in February or early March, when the plant is growing strongly and it is possible to see buds of new growth forming.

Remove all the dead and weak growth and then cut back all stems to the healthiest pair of buds. Tie in branches securely, spacing them evenly over the support so that there is enough room for the new shoots and their crop of flowers; blossoms should not overlap, nor should the stems be tied in so tightly that they are likely to be damaged as they grow.

Group three contains the late flowering hybrids which produce flowers on new growth, that is growth made during the current season. This group includes the *Jackmanii*, *viticella*, *texensis* and

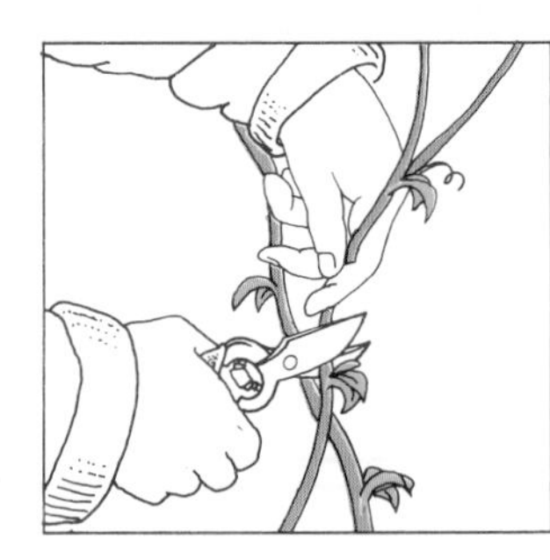

When pruning early-flowering hybrids in spring, cut back old wood to a healthy pair of buds.

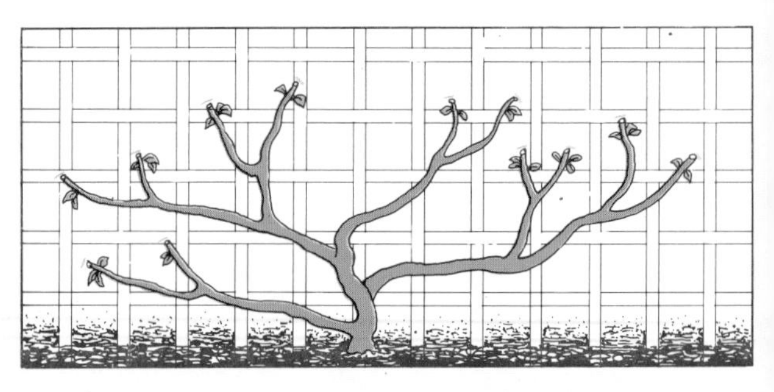

Remove dead and straggly growth. Then neatly tie in the stems to the support, spreading the branches evenly.

The species clematis, such as *C. macropetala*, do not require pruning. But after a few years they may become overgrown and should be clipped over to tidy them up.

their cultivated forms and hybrids, and all the large-flowered, late summer flowering hybrids. The clematis in this group begin flowering in July and usually have two or more flowers on each stem. With these clematis all the old growth dies down naturally each year, so it is quite an easy job to prune hard and remove all the dead growth. This can be done anytime between January and March, although young buds spurred into growth by the pruning may be damaged by winter frosts if an early time is chosen.

Cut the stems right back to the bottom pair of buds, about 15-30cm (6-12in) from the ground. These plants readily produce new shoots from below ground and hard annual pruning encourages this so that, in effect, you are beginning each year with a healthy new plant. Herbaceous types of clematis such as *recta* and *heracleifolia* should also be pruned in this fashion.

Additionally, there are clematis on which pruning is optional; they will produce abundant flowers on old or new wood, although the flowers produced early in the year on old wood (on a plant pruned as for group two) will be larger than those produced during late summer (on a plant pruned as in the third group). Clematis in this category include 'Henry I', 'Ville de Lyon' and 'William Kennet', and other clematis that produce new growth while flowering and then continue to flower on both the old and new wood. If you have a clematis and are uncertain about its category, note when it comes into flower and then study its habit of growth during the summer months.

If you have a clematis that has not been properly pruned and has become a choked and knotted mass, or a vast stretch of naked stem supporting an inadequate cluster of blooms well out of sight, do not be tempted to take drastic measures and hard prune it to the ground. Clear the jumble gradually over a season or two, and in the case of a bare stem, prune only as far down as the healthiest pair of buds. This may mean that the flowers will be higher than desired, but this can be remedied by training new long shoots downwards to cover the plant's bare legs.

# AFTERCARE

Once the clematis is established you must see to it that the plant is regularly fed and watered. The most common mistake made with clematis once it is planted is neglecting to feed it. Even if the soil has been properly prepared and fertilized before planting, the quality and quantity of flowers will suffer without additional feeding.

During spring, when the soil is moist from rain, fork a handful of sulphate of potash into the soil around the plant. Then spread a surface mulch of well-rotted manure or compost at least 8cm (3in) deep around the base of the plant; take care that it does not actually touch the base of the stem. The mulch will help to conserve moisture during dry spells and also act as a regular feed as it is taken into the soil. If no compost or manure is available, use a mixture of one handful of 'Forest Bark' Ground and Composted Bark to one bucket of peat.

Ideally, clematis should receive at least two gallons of water several times a week during the summer, but this will be influenced by the weather conditions – if it is exceptionally dry you may have to give it more water – and by the type of soil. Whereas heavy soil will retain the moisture better, a clematis planted in a loose sandy soil will require a great deal more water.

All the late spring-and-summer-flowering types should be given a liquid feed, such as ICI Liquid Growmore, once a week during the growing season, before flowering. If this is neglected, the flowers will be smaller and fewer in number each year. The early-flowering species of clematis are not so demanding, but will benefit from an annual application of a general low-nitrogen fertilizer such as bonemeal, anytime between December and February.

**Moving established clematis** Nearly every gardener is tempted to alter planting schemes sooner or later, and this means knowing how to set about lifting mature established plants without harming them in the process. Hybrids, because of their tougher root systems, are more likely to survive the move than the fibrous-rooted species.

There are three points to remem-

Each spring, apply a generous mulch to the base of the clematis. Then water liberally at regular intervals throughout the growing season.

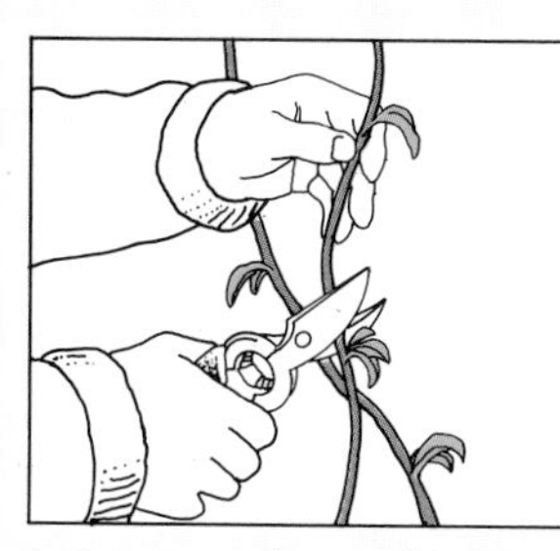

1. Before moving an old plant to a new site, prune back to healthy buds.

2. Dig out the plant at least 60-90cm from the base of the stem.

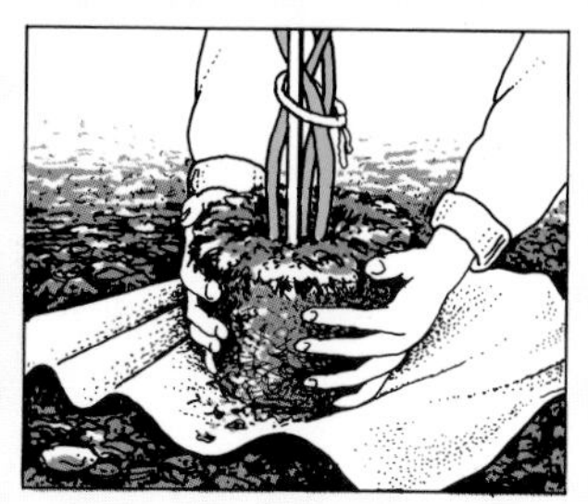

3. Wrap the root ball in an old sack.

ber: do not move a plant except during its dormant period, ideally late January to February; take enough of the old root system to support the plant while it re-establishes itself; and, in the case of a climbing plant such as clematis, do not cut back into the old wood. Find the strongest growing pair of leaf axil buds and make the pruning cut just above them, otherwise there will be no new shoots the following season.

Position a strong cane next to the main stem, taking care not to damage the root system by poking the cane into the ground too close to the base of the stem – about 15cm (6in) away is adequate. Tie the stem to the cane and then begin the excavation. With the edge of the spade, mark out a line about 60-90cm (2-3ft) around the base of the plant. Dig out the root ball to a spade's depth, working the end of the spade well under the plant to loosen it. Now carefully lift the plant by the root ball on to a heavy polythene sheet or old sack; depending on its size, you may need someone to help you. Do not be tempted at any time during the moving to loosen the plant by pulling its stem or lifting it by the stem.

*Clematis armandii* (left), a vigorous evergreen species, flowers from January to May. If growth is to be restricted, remove all flowered shoots when blooming is over. Train subsequent growths.

# PROPAGATION

Clematis, both species and hybrids, are easy to propagate and there are a number of good reasons for knowing how to do it. If a hybrid wilts or is chopped down by an inattentive gardener, you will have another to replace it; or if, when visiting a friend's garden, you spy a choice clematis that you would dearly love in your own garden, you can ask for a cutting.

**Raising from seed** Species clematis will usually seed themselves quite readily and young plants in the border, or wherever they happen to be growing, can be lifted and bestowed on gardening friends or the church bazaar. Hybrids raised from seeds are rather hit and miss; whereas a species will generally come true to type, a hybrid seedling can develop into something quite unlike its parent. So unless you relish a surprise that might be disappointing, it is not really worth raising a hybrid seedling.

If you decide to gather and raise seedlings of the early flowering species, begin in September as the seed should have ripened by then. Remove the silky seed tails and sow the seed 5mm (⅛in) deep in a proprietary seed compost, such as 'Kericompost', in seed trays or pots. Cover the trays and put them in a cool, shaded place in the greenhouse or coldframe. Do not allow the compost to dry out.

The seed should have germinated by the following spring. As soon as it has sprouted, remove the cover and expose to increasing doses of sunlight. Pot-on individually when the seedlings are large enough to handle, and within a year they should be ready to plant in the garden.

**Plants from cuttings** This is a fairly tricky business and chances of success are increased if you have a bottom-heated propagator, or at the very least a DIY propagating unit. This is simply a tray with a straight-sided vented cover, available in a range of sizes.

Cuttings should be taken of soft growth during May, June or July and consist of one pair of leaves and at least 5cm (2in) of stem. Cut away all growth above the leaves and then remove the leaf that has a twisted

**Cuttings** 1. Take soft-wood cuttings in spring.

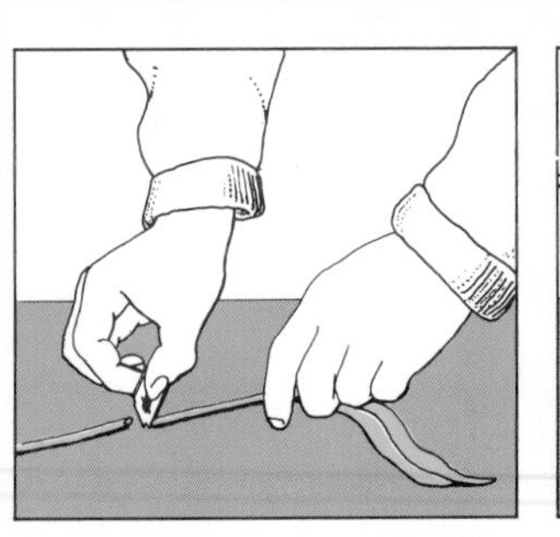

2. Trim away the leaf with the curly stem.

3. Dip the ends in hormone rooting powder and place several around the edge of a pot.

stem. Trim the bottom of the stem diagonally. If the leaves are large, which they may be on a cutting taken from a mature plant, trim these by half. The neatest trims are made with a razor blade against a flat surface.

Dip the ends of the cuttings into a rooting powder, such as 'Keriroot', and then push into trays of compost, making sure that the leaves do not overlap or touch the compost or cover. Water in with a solution of Benlate + 'Activex'. Cover the propagator and put in a heavily shaded place. When you notice signs of growth, gradually admit air and light. Continue to spray with Benlate + 'Activex' but do not over-wet the compost. Cuttings from vigorous species will probably have rooted by the end of August, but those from large-flowered hybrids will take longer, so leave them until at least March the following year before potting on individually.

**Layering** This is not only the surest way to succeed, it is also the easiest way to produce new clematis plants from old. Begin at the end of summer when the stems have had a chance to mature.

Fill a 10cm (4in) flowerpot with potting compost and sink it in the soil near to the base of a parent plant so that its rim is level with the soil. Then choose a sturdy stem, detach it from the support and gently bend it down to the flowerpot,

Many species clematis are easily grown from seed; *C. macropetala* (above and left) will self-sow readily.

centring a healthy pair of leaves on the surface of the compost. Peg it in place with a bend of wire or a wooden clothes peg. It is a good idea to split the stem carefully below the leaf node and dip it in hormone rooting powder before pegging down.

Attach the growing end of the shoot to a small cane inserted near the peg and then firm a layer of soil over the top of the pot. Keep moist and do not disturb until the following March when you should sever the new plant from the parent and lift the pot from the ground. Plant and prune it exactly as you would a new plant.

**Division** Old, established plants of the large-flowered hybrids will produce a number of new shoots each year. Each shoot will have its own roots and can be detached from its neighbours and planted out as a new plant. Herbaceous varieties can be lifted and divided during the dormant season just as you would any other herbaceous garden plant.

**Layering**
1. Stick pot in soil next to base of main stem. Fill with compost.

2. Make small slit in new shoot, bend to pot surface and peg to soil.

*C. viticella rubra* (right) is a vigorous climbing species. Take stem cuttings in July or sow seeds in October.

# PESTS AND DISEASES

Many gardeners are discouraged from growing the large-flowered hybrids by terrible tales of clematis wilt. Certainly it is discouraging to see an apparently healthy plant simply keel over and die, but if this happens, chances are new shoots will appear within a few weeks if a few simple remedial steps are taken. You are more likely to succeed in this if the clematis was planted deeply to begin with, with a part of the stem below ground.

First, cut down the dead growth to the ground and burn it. Then ensure that the roots are kept moist, cool and nourished; give a liquid feed once a week. Within several weeks you should see new growth emerging, but do not despair if nothing happens as occasionally a clematis can go into a dormant period.

Once growth appears, begin a regular programme of spraying with a fungicide such as Benlate + 'Activex', using this from spring through to autumn. Drench the base of the main stem and spray the surrounding soil. Clematis that have a regular and substantial supply of water are less prone to wilt. So if you garden on dry, quickly draining soil it is important to ensure that your plants never go thirsty.

*Clematis* **'Niobe'**

Sometimes a plant will suffer from wilt several years running, and while some authorities recommend taking out the plant, rinsing all the soil from its roots and soaking it in Benlate + 'Activex' for a day or two before planting in a new site, this disturbance can be detrimental. It is better to persevere with cutting it down each time, as the plant will in all likelihood grow out of its 'wilt phase'.

Slugs and snails are horrid pests, diligent in their pursuit of tender shoots. The only way to thwart them is to be equally diligent in your application of slug pellets used according to the manufacturer's instructions. Alternatively, a ring of coal ash or limy grit put around the stem – about 4cm (1½in) away from it – might deter slugs and snails; they don't like the scratchiness. Tiny black slugs that live below the soil surface can be killed by watering in a liquid control, but you will have to apply it regularly until the shoots are well grown.

Early flowering species are sometimes disfigured by earwigs chewing large holes in leaf and petal – young plants can loose their entire show of flowers, and that is very disappointing. Spray regularly with 'Sybol' an insecticide designed to combat this sort of pest. This should be done late in the evening so the pesticide solution does not evaporate and the foliage is still wet when the earwigs come out to feed. Make sure you douse the dark shady spots where earwigs lurk during the day out of the sunlight.

# PLANTING FOR A PURPOSE

Usually clematis are simply trained against supports on walls or fences, but they are versatile plants and can be used in a number of other ways in a garden setting.

**Ground cover** Vita Sackville-West, the creator of the beautiful gardens of Sissinghurst, recommended growing clematis over a square of garden netting, pegged horizontally over a section of the border so that the flowers would make a low blanket to be viewed from above. While this idea has its attractions, it would require a fairly large section of border to be effective. But clematis can be grown as ground cover where space is at a premium, through low-spreading shrubs and border plants.

There are advantages to planting clematis with low-growing shrubs: the foliage provides the shade clematis like at their roots and gives some protection against frost for early-flowering varieties. A beautiful floral display can also be achieved by combining plants.

Dwarf evergreens and winter flowering heathers can appear dull during summer, but make an excellent background for the summer-flowering hybrid clematis. The *viticella* group is a good choice for this purpose, providing flowers throughout the summer and into early autumn. Allow the slower evergreens to establish themselves before planting the clematis in such a situation, otherwise the clematis will compete greedily for the available food in the soil.

Naturally colour should play a major role in your selection. Think about allowing some of the darker-flowered hybrids to grow through grey-leaved shrubs such as senecio; most of the rich reddish purples would look well with the yellow flowers of this shrub, or choose a clematis whose flowering is over by the time the shrub begins to bloom. Other shrubs, such as lavender, cistus, rock roses, hebe and artemisia are all suitable hosts. Many of the evergreen shrubs such as prunus, sarcococca, *Daphne odora* and other shrubs that do well in semi-shaded spots, are good hosts for large-flowered hybrids such as 'Nelly Moser', whose flower colour would fade in sunlight.

*C. venosa violacea* growing through a support of evergreens (right); the shrub provides shade for the roots of the clematis, and the flower colour is perfectly complemented by the green foliage. Clematis *montana* 'Rubens' (below) will, in three or four years, form a dense thicket wreathed in starry flowers during early summer.

For true ground cover (a plant that is dense enough to completely hide the soil) the nearest you will get are some of the more rampant species such as *macropetala* and the *montanas*. These do nicely scrambling over banks at driveways or as a protective blanket over a bed of summer-flowering lilies, which also like having their feet in the shade and heads in the sun.

**Through trees and shrubs** It is an old-fashioned, cottage garden tradition to train clematis into fruit trees and evergreens, and anyone who has seen 'Etoile Violette' garlanding a ripening apple tree might be taken with this idea.

It is relatively easy to grow a clematis into a tree or large shrub successfully as long as you remember to position the clematis far enough from the base of the host so that it doesn't have to compete for food and water, and as long as it has enough space to produce its flowers easily. A large-flowered hybrid will not have space to open its blooms fully among a dense crowd of branches, so a light pruning would be beneficial. Also, clematis will grow towards the light to flower, so again the removal of some branches to increase the area of light will help to make a better display.

Generally, it is best to use the late-flowering hybrids to grow into trees and shrubs as these can be

'Margot Koster' (above) is a good subject to train on supports. Bare lower stems, like on this *C. jackmanii* (right) can be concealed by tall subjects.

gently pulled out of the host when the leaves of both begin to die back, and the faded clematis growth is cut back. Proper pruning can then be done in spring and new growth trained into the host.

Early-flowering hybrids can be used to grow over, rather than through, compact shrubs such as olearia, pyracantha, hardy fuchsia, and shrubby honeysuckles like *Lonicera nitida*. Lilac, viburnum, cotinus, flowering cherry, rhododendrons, malus, mahonia and laburnum are some of the suitable host plants for either type of clematis.

**With other wall-trained plants**
The most familiar way of growing clematis – against a wall or a fence – can be varied delightfully by associating them with other climbers. Non-flowering climbers, like the various ivies, or those with inconspicuous flowers such as *Lonicera japonica* 'Aureoreticulata' are given a new dimension with the addition of a large flowered hybrid. Alternatively the choice of host plant and clematis can be decided by colour schemes, so that both plants flowering together are complementary, or by contrasting foliage of host with flower of clematis: imagine the pink-tinged variegated leaves of *Actinidia kolomikta* with clematis 'Ville de Lyon'. The variations are endless, the only criteria being that the host is at least three years old before a clematis is introduced, that the soil and light requirements are compatible and that the pruning requirements are basically the same.

For example, if you choose to train an early-flowering hybrid that requires light pruning in spring through a climbing rose that needs extensive pruning at that time of year, the task would be nerve-racking. To avoid damaging the clematis use a late-flowering clematis that is pruned hard at the same time as the rose. The pruning rule also applies when growing several clematis together – make sure they all have the same needs and you won't have to worry about disentangling them.

**Container growing** It is best to use early flowering, large-flowered hybrids for this purpose, as these tend to have the most compact habit and are easiest to train into a tidy shape.

As with any container-grown plant, you must be sure to use a container that is large enough for the plant, and to see that there is adequate drainage. Place a layer of broken crocks or fine gravel in the bottom of the container and fill to within 5cm (2in) of the top. Use a purpose-made compost, such as the soil-based John Innes Composts. Plant the clematis as you would in the open garden, planting it deeply to encourage shoots from below ground.

In early spring prune the plant hard to within 25cm (10in) of the soil, making the cut just above a healthy pair of buds. Several new stems will grow from the buds and these should be left to grow on and then pinched out after three more buds have formed on each. The new stems that will grow from these should then be tied in carefully to four or more canes placed around the edge of the container. Bend the stems diagonally and tie them to the nearest canes so that they spiral around them and are held almost diagonally. This will allow enough room for the buds to mature.

The clematis can now be trained into a wall support by inclining a cane or two from the container to the wall and leading the shoots of the clematis upwards. Alternatively, if the clematis is being grown as a free-standing specimen, you can train it over a wigwam of slender canes or use a purpose-made support, such as the type used to train weeping roses.

Container-grown clematis can be used as an attractive feature in the garden.

As the plant grows, tie in all new growth to keep it neat. Water and feed as you would in the open garden, but remember that containers dry out more quickly. When watering, be generous and make sure the soil is thoroughly drenched.

After flowering, let the plant die back and then begin pruning again in spring, cutting back only as far as the largest, healthiest pair of buds. Large-flowered, late summer flowering hybrids may also be trained in this manner, but it is more difficult to get them into a nice compact shape. Prune in the beginning, as with any early flowering hybrids, then the following season prune all growth back almost to the soil level or the bottom pair of healthy buds. New growth, as it appears, should be tied in the spiralling manner before it is 30cm (1ft) long.

# THE CLEMATIS YEAR

### JANUARY

Now is the time to move established clematis to new positions. Lift and divide herbaceous clematis.

### FEBRUARY

Prune all newly planted clematis down to 30cm (1ft) from the ground. Established clematis can still be moved. Check mulches around plants and replace as necessary. Begin light pruning of early flowering clematis hybrids. Cut stems back to strong pair of buds. Remove all other dead or weak growth. Begin hard pruning of the late-flowering hybrids. Cut stems back to 15-30cm (6-12in), to bottom pair of healthy buds. After pruning, check all supports and repair where necessary. Tie in main stems. Complete division of herbaceous clematis.

### MARCH

Complete all pruning. Begin feeding as new growth appears. Apply mulches to base of plant, first making sure that soil is moist. New clematis may be planted now in site prepared the previous autumn. Be sure to water well until plant is established.

### APRIL

Continue feeding regularly. Use a liquid feed for best results. Take precautions against slugs, snails and other pests that would be likely to destroy new shoots. Tie in new stems to supports.

### MAY

Carry on feeding and be sure plants have an adequate supply of water as this is an important growth period. Container-grown plants may still be planted. New plants may be raised by layering.

Clematis 'Marie Boisselot and 'W. E. Gladstone'

## JUNE

After flowering is finished on the spring-flowering clematis, clip them over to remove all weak and dead stems and generally tidy them up. Layering may still be done to raise new stocks. Plants showing signs of wilt should be dealt with immediately. Do not relent in keeping the clematis fed and watered.

## JULY

Many of the hybrids will now be in flower, while some may be resting between shows. The only thing required now is to make sure they are watered and, if a liquid feed is being used, that this is also applied on a regular basis. Choose new clematis for planting next month. Begin preparing the site for planting new clematis.

## AUGUST

Complete preparations of site for new clematis. Anytime between the end of this month and mid-November is the ideal time to plant new clematis.

## SEPTEMBER

Gather seed for sowing to raise new plants of species clematis.

## OCTOBER

Continue to gather seed until this task is completed.

## NOVEMBER

New clematis should be planted by the middle of this month.

## DECEMBER

See that all clematis are secure against strong winter winds. Check supports of species clematis, especially if the plants are large.

# SIXTY OF THE BEST

Here is a guide to 60 of the most popular clematis, commonly available from garden centres or specialist nurseries. Use this list to help you choose plants for colour, size of flower, or site (is it suitable to grow through hedges, near trees or in containers?). When you have chosen your clematis, you can refer to this list as a guide to pruning requirements.

**ALPINA FRANCES RIVIS**
This variety has the best blue flowers of the species *alpina*. They are small nodding, bell-shaped deep blue flowers with a cluster of petal-like white stamens in the centre.
**Flower size:** up to 5cm (2in).
**Height:** up to 2.5m (8ft).
**Blooms:** April and May. Suitable for any aspect.
No pruning necessary.

**BARBARA JACKMAN**
Good for container growing or for training through shrubs. The flowers are soft, blue-mauve with pronounced crimson bars; the blue fades but the bars retain their colour. Stamens are off-white.
**Flower size:** up to 15cm (6in).
**Height:** up to 3m (10ft).
**Blooms:** May and June with a second show in August. Suitable for any position, but flowers retain their colour best in partial shade.
Prune lightly in early spring.

**BEAUTY OF WORCESTER**
A hybrid with double flowers, coloured rich blue with a tinge of red; creamy white stamens. Because of its neat, compact growth it is ideal for container growing.
**Flower size:** 15-20cm (6-8in).
**Height:** up to 3m (10ft).
**Blooms:** May and June with a second show of single flowers in late summer. Prefers a sheltered position.
Prune lightly in early spring.

**BEES JUBILEE**
A very free-flowering improvement on 'Nelly Moser', this hybrid has deep lavender-pink flowers with a dark crimson bar.
**Flower size:** 15-20cm (6-8in).
**Height:** up to 2.5m (8ft).
**Blooms:** May and June with a substantial second show in late summer. Suitable for any position, but flower colour is best in partial shade.
Prune lightly in early spring.

***Alpina* 'Francis Rivis'**

**'Barbara Jackman'**

'Bees Jubilee'

## CAPTAIN THUILLEAUX

A particularly eye-catching hybrid because of the broad striping of the petals. Similar to the above, but with a broader rosy pink bar against a pale background.
**Flower size:** to 20cm (8in).
**Height:** up to 3m (10ft).
**Blooms:** May and June, and again in late summer. Suitable for any aspect.
Prune lightly in early spring.

## CIRRHOSA BALEARICA

This clematis is valuable as a screen because the dark shiny leaves are evergreen. The nodding, bell-shaped creamy yellow flowers have a faint perfume; the foliage is finely cut.
**Flower size:** to 2.5cm (1in).
**Height:** up to 3m (10ft).
**Blooms:** January to March. Suitable for any position, but does best in shelter if there is danger of frost.
No pruning necessary.

'Captain Thuilleaux'

*C. cirrhosa balearica*

## COMTESSE DE BOUCHAUD

A vigorous grower, but not to a great height. Abundant flowers are largest at start of season, becoming smaller as it progresses. They are lavender pink with neatly rounded petals and creamy stamens.
**Flower size:** to 15cm (6in).
**Height:** up to 4m (13ft).
**Blooms:** Continuously from June to September. Suitable for any aspect.
Prune hard in spring.

## COUNTESS OF LOVELACE

A hybrid with pale lilac-blue double flowers, followed by a small show of single flowers later in the season. It can be slow to establish itself, but patience will be rewarded with a vigorous, free-flowering plant.
**Flower size:** to 15cm (6in).
**Height:** up to 3m (10ft).
**Blooms:** June to September. Does best in a sheltered position.
Prune lightly in spring.

'Daniel Deronda'

'Countess of Lovelace'

'Comtesse de Bouchaud'

**DANIEL DERONDA**
This plant has pale violet flowers with contrasting yellow stamens. The first flowers are semi-double, becoming single later in season.
**Flower size:** to 20cm (8in).
**Height:** up to 3m (10ft).
**Blooms:** June to October. Suitable in any warm, sunny position.

**DOCTOR RUPPEL**
One of the showiest large-flowered hybrids. The flowers are broad bars of deep pink, edged with pale rose; large cluster of gold stamens.
**Flower size:** to 20cm (8in).
**Height:** up to 4m (13ft).
**Blooms:** May, June and September. Good for any position in the garden. Prune lightly in the spring.

**EDITH**
A hybrid species that is valuable for its long and abundant flowering. It has white flowers, tinged pale lavender at the edges with dark reddish stamens.
**Flower size:** to 20cm (8in).
**Height:** up to 3m (10ft).
**Blooms:** A long season, often from May to September. Suitable for any position.
Prune lightly in spring.

**'Edith'**

**ELSA SPATH (syn. XERXES)**
A very free-flowering hybrid. The deep lavender blue petals contrast with the ruby coloured anthers. The first flowers have large, limp petals; the later ones are smaller.
**Flower size:** to 20cm (8in).
**Height:** up to 3m (10ft).
**Blooms:** May to September. Flowers best in south, east or west-facing position.
Prune lightly in spring.

**ERNEST MARKHAM**
One of the best reddish colourings among the clematis. The flowers are a fine reddish-purple with golden brown anthers.
**Flower size:** to 15cm (6in).
**Height:** up to 3.5m (11½ft).
**Blooms:** Continuously from July to October. Plant in east, south or west-facing site.
Prune hard in spring. A vigorous grower, it can be left unpruned to flower on old wood.

**'Ernest Markham'**

**ETOILE VIOLETTE**
A vigorous and very free-flowering hybrid. It has rich, deep purple flowers with bright yellow stamens.
**Flower size:** to 10cm (4in).
**Height:** up to 3.5m (11½ft).
**Blooms:** Throughout July and August. Situate in south, east or west-facing position to preserve colour.
Prune hard in spring.

**FLORIDA BICOLOR**
(syn. **SEIBOLDII**)
One of the climbing varieties of *c.florida*. Flowers are white with a central boss of stamens, beginning green maturing to purple. The unusual blossoms look like those of a passion flower; the star-like flowers look well against dark foliage.
**Flower size:** to 10cm (4in).
**Height:** up to 2.5-3m (8-10ft).
**Blooms:** From late June to September. Requires a sheltered south or west-facing position.
Prune hard in spring.

'Etoile Violette'

**GENERAL SIKORSKI**
A hybrid that can be relied upon to produce an abundance of neatly shaped flowers of good colour over a long season. Flowers are lavender blue with lemon yellow stamens.
**Flower size:** to 15cm (6in).
**Height:** up to 3.5m (11½ft).
**Blooms:** From June to September. Don't plant in north-facing position.
Prune lightly in spring.

**HAGLEY HYBRID**
A good plant for containers as it is tidy and compact and produces masses of flowers over a long summer season. The flowers are a rosy lavender colour with brownish purple anthers.
**Flower size:** to 10cm (4in).
**Height:** up to 2.5.m (8 ft).
**Blooms:** Continuously from June to August and September. Position out of direct sun; good for north-facing aspect.
Prune hard in spring.

'Hagley Hybrid'

'Henryi'

## HENRYI

One of the oldest hybrids, very good for cut flowers. They are creamy white, set off by dark brown stamens.
**Flower size:** to 10cm (4in).
**Height:** up to 3m (10ft).
**Blooms:** June to August-September. Suitable for any position.
Prune hard in early spring; can be left unpruned.

## H.F. YOUNG

Tiny, compact habit, free-flowering in one of the best blue shades – duck-egg blue flowers, cream stamens.
**Flower size:** to 15cm (6in).
**Height:** up to 3m (10ft).
**Blooms:** May and June with a second show in late summer. Not suited to a north-facing aspect.
Prune lightly in spring.

'H. F. Young'

'Horn of Plenty'

## HORN OF PLENTY

A free-flowering and vigorous hybrid that can be used to grow through trees and cover large areas of wall. Flowers are deep lavender rose that fades to soft mauve. Prominent boss of deep reddish purple stamens.
**Flower size:** to 15cm (6in).
**Height:** to 2.5m (8ft).
**Blooms:** July to September. Suited to any position.
Prune hard in spring.

**JACKMANII ALBA**
Not a particularly vigorous hybrid, but interesting for its curious flowers. They are white with a bluish tinge on the first show of semi-double flowers; later flowers are single.
**Flower size:** to 10cm (4in).
**Height:** up to 2.5m (8ft).
**Blooms:** Continuously from June to September. Suited to any position. Prune lightly to permit early show of flowers, or hard-prune if desired.

**JACKMANII SUPERBA**
This is the most popular and widely grown clematis, loved for the rich colour of its abundant flowers. They are dark velvety purple with green stamens.
**Flower size:** to 15cm (6in).
**Height:** up to 4m (13ft).
**Blooms:** Continuously from June to September. Good for any position. Prune hard in spring.

'Jackmanii Superba'

**JOHN WARREN**
The strikingly marked, large flowers are freely produced over a long summer season. They are deep rose pink bars, edged with soft pinky grey that fades to pale lilac pink; attractive brown stamens.
**Flower size:** to 15cm (6in).
**Height:** up to 3m (10ft).
**Blooms:** June to September. Flowers retain colour best out of direct sun. Good for a north-facing aspect. Prune lightly in spring.

**KATHLEEN DUNFORD**
A hybrid with unique flowers: unlike other doubles, the petals of each layer are the same length. The semi-double flowers are rose mauve with golden stamens.
**Flower size:** to 15cm (6in).
**Height:** 3m (10ft).
**Blooms:** May and June with a second show in late summer. Suitable for any aspect.
Prune lightly in spring.

**LADY BETTY BALFOUR**
One of the finest late-flowering clematis; plant in full sun for the best flowers. Very vigorous, with deep purple flowers that fade to soft blue and a white bar on reverse of petals.
**Flower size:** to 15cm (6in).
**Height:** over 4m (13ft).
**Blooms:** Late August to October. A warm east, south or west-facing aspect is best.
Prune hard in spring.

**LADY CAROLINE NEVILL**
A moderately vigorous hybrid with lavender blue flowers and pleasing beige anthers.
**Flower size:** to 18cm (7in).
**Height:** up to 3m (10ft).
**Blooms:** June to August. Not suitable for north-facing aspect.
Prune hard in spring. If pruned lightly in spring will produce semi-

double flowers on old wood, single flowers on new wood.

## LASURSTERN

A fairly vigorous clematis that has a good first crop of large showy flowers; the second show has smaller flowers. They are deep, rich lavender blue that fades to duck egg blue. The reverse of the petals is marked with a pale green bar.
**Flower size:** over 15cm (6in).
**Height:** up to 3m (10ft).
**Blooms:** May, June with second show in September. Suitable for any position. Prune lightly in spring.

'John Warren'

'Kathleen Dunford'

'Lady Betty Balfour'

'Lasurstern'

**LOUISE ROWE**
A new clematis and one of the most beautiful, having double, semi-double and single flowers simultaneously. They are palest lavender with a cluster of lemon-coloured stamens.
**Flower size:** to 15cm (6in).
**Height:** up to 2.5m (8ft).
**Blooms:** June and July with second show in September. Give a south, west or east-facing position. Prune lightly in spring.

**MACROPETALA**
This species is reliably hardy and free-flowering. It also has attractive, fluffy seedheads and tiny, nodding blue flowers with white stamens.
**Flower size:** to 5cm (2in).
**Height:** 2.5m (8ft).
**Blooms:** April and May. Suitable for any position. No pruning required.

**MACROPETALA MARKHAMII** (syn. **MARKHAM'S PINK**)
A pink form of the above, with greenish stamens. It has the same attributes as the blue form, except that the rosy petals of 'Markhamii' are edged with paler tint of lilac, making it more interesting.
**Flower size:** to 5cm (2in).
**Height:** up to 3m (10ft).
**Blooms:** April and May. Suitable for any position. No pruning needed.

**MADAME BARON VEILLARD**
A good, late-flowering clematis that is usefully vigorous and can cover a large wall or other support in one season. Lilac rose flowers with pale green stamens.
**Flower size:** to 10cm (4in).
**Height:** up to 3.5m (11½ft).
**Blooms:** September and October. Requires a warm south-, east- or west-facing position. Prune hard in spring.

**MADAME EDOUARD ANDRÉ**
This hybrid produces an abundance of deliciously coloured flowers over a long season. They are rich claret red with creamy stamens.
**Flower size:** to 10cm (4in).
**Height:** up to 2.5m (8ft).
**Blooms:** June to September continuously. Plant in any position. Prune hard in spring.

**MARGOT KOSTER**
Vigorous and free-flowering, the blossoms have an open floppy appearance. They are a cheerful rosy-mauve with widely spaced petals.
**Flower size:** to 10cm (4in).
**Height:** up to 2.5m (8ft).
**Blooms:** Throughout July and August. Will do well in any position. Prune hard in spring.

**MARIE BOISSELOT**
(syn. **MADAME LE COULTRE**)
One of the best white clematis, prolific and vigorous, it also has attractive foliage. The flowers have a pinkish tinge when opening but soon turn clear white. The stamens are bright yellow.
**Flower size:** to 15cm (6in).
**Height:** up to 4m (13ft).
**Blooms:** June to the end of September. Suitable for any aspect. Prune lightly in spring.

C. macropetala 'Markham's Pink'

'Madame Baron Veillard'
*C. macropetala* (left)

**MAUREEN**
A neat, rather bushy plant, free-flowering and valuable for its velvety rich royal purple colouring.
**Flower size:** to 15cm (6in).
**Height:** up to 2.5m (8ft).
**Blooms:** June to September. Plant in east, south or west-facing site. Prune lightly in spring.

**MONTANA ELIZABETH**
A fast-growing and free-flowering variety, it will soon cover fences or screen sheds and is useful for growing through trees. Flowers are soft shell-pink with a very sweet scent.
**Flower size:** to 2.5-5cm (1-2in).
**Height:** up to 9m (30ft).
**Blooms:** May-June. Does well in any position. No pruning necessary.

**MONTANA ODORATA**
An old-fashioned favourite, it has sweetly scented white flowers. All other characteristics are as for *m.* 'Elizabeth.'

***Montana* 'Elizabeth'**

**MONTANA RUBENS**
A particularly attractive montana, its delicate rosy pink flowers and yellow stamens are nicely set off by the warm bronze tone of the foliage.
**Flower size:** to 5cm (2in).
**Height:** up to 8m (26ft).
**Blooms:** May and June. Suitable for any position. No pruning needed.

**MONTANA TETRAROSE**
Rosy lilac, with bronze foliage. Shares all the characteristics of the *montana* group, except that the rosy lilac flowers are the largest.

**MRS CHOLMONDELEY**
Vigorous and free-flowering over a long season, this is one of the old favourite hybrids. Flowers are a fine lavender blue with darker veining.
**Flower size:** to 15cm (6in).
**Height:** up to 3m (10ft).
**Blooms:** May to September. Suitable for any position in the garden. Prune lightly in spring.

***Montana* 'Rubens'**

'Mrs Cholmondeley'

*Montana* **'Tetrarose'**

## MRS GEORGE JACKMAN

A fairly vigorous hybrid with soft, off-white flowers and pale brownish stamens. Early flowers are semi-double, later ones single.
**Flower size:** 15cm (6in).
**Height:** up to 2.5m (8ft).
**Blooms:** May and June with a second show in August-September. Suitable for any position in the garden. Prune lightly in spring.

## MRS SPENCER CASTLE

The early flowers of this clematis are fully double, the later ones are single. They are rosy heliotrope in colour, with golden yellow stamens.
**Flower size:** to 10cm (4in).
**Height:** up to 2.5m (8ft).
**Blooms:** May and June with a second show in September. Suitable for any position. Prune lightly in spring.

## NELLY MOSER

Most people's idea of the perfect hybrid clematis, this is among the five top sellers at garden centres. A strong-grower that can be trained through trees and shrubs, providing it has the shade needed to preserve its colour. It has pale mauve flowers with a deep carmine red bar down the centre of each petal.
**Flower size:** to 15cm (6in).
**Height:** up to 3.5m (11½ft)
**Blooms:** May and June and again in September. Flowers fade in sunlight, so give a sheltered east, west or north-facing site. Prune lightly in spring.

'Nelly Moser'

'Niobe'

'Perle d'Azur'

'Proteus'

'Richard Pennell'

## NIOBE

A strong-growing hybrid especially valued for its superb colour – dark claret red with a sheen of velvet, set off by greenish-gold stamens.
**Flower size:** to 15cm (6in).
**Height:** up to 3m (10ft).
**Blooms:** June to September. Do not plant in a north-facing position. Prune hard in spring.

## PERLE D'AZUR

A strong-growing and free-flowering hybrid that will quickly grow into trees and shrubs; also makes good ground cover. Azure blue with green stamens.
**Flower size:** to 10cm (4in).
**Height:** to 3.5m (11½ft).
**Blooms:** July to September. Suitable for any position. Prune hard in spring.

## PROTEUS

A moderately vigorous hybrid, useful in containers. Flowers are mauve pink with golden yellow stamens. Wonderful double flowers in early summer, followed by a second show of single blossoms.
**Flower size:** to 15cm (6in).
**Height:** up to 3m (10ft).
**Blooms:** May and June, August and September. Give a warm south-, east- or west-facing position. Prune lightly in spring.

## RICHARD PENNELL

One of the newer hybrids and unusually beautiful, with deep lavender rose flowers, set off by strikingly coloured anthers and swirling stamens. It can be trained through shrubs and grown in containers.
**Flower size:** over 15cm (6in).
**Height:** up to 3m (10ft).
**Blooms:** June to September. Suitable for any position. Prune lightly in spring.

### ROUGE CARDINALE

Valuable for its unusual colouring – rich magenta red with brown stamens. This hybrid is free-flowering over a long season.
**Flower size:** to 15cm (6in).
**Height:** up to 3m (10ft).
**Blooms:** June to September. Suitable for any aspect. Prune lightly in spring.

### SILVER MOON

Vigorous, free-flowering and rather bushy plants, ideal for containers. Flowers are pearly grey-white with yellow stamens.
**Flower size:** to 15cm (6in).
**Height:** up to 3m (10ft).
**Blooms:** June to September. Suitable for any position, but its ghostly pallor is best seen in a shaded north-facing site. Prune lightly in spring.

### STAR OF INDIA

Free-flowering, the individual blossoms are nicely shaped; they are dark purple in colour with a carmine bar. Can be used in tall trees or as ground cover.
**Flower size:** to 10cm (4in).
**Height:** to 3.5m (11½ft).
**Blooms:** Continuously from June to September. Suitable for any position. Prune hard in spring.

### TEXENSIS GRAVETYE BEAUTY

Vigorous, reliable and free-flowering; this variety of *texensis* makes good ground cover. It has ruby red, bell-shaped flowers that gradually open out to a star shape.
**Flower size:** to 5cm (2in).
**Height:** up to 2m (6½ft).
**Blooms:** July to September. Good in any position. Prune hard in spring.

'Rouge Cardinale'

'Silver Moon'

***Texensis* 'Gravetye Beauty'**

**'Star of India'**

**THE PRESIDENT**
Another of the top five clematis, deservedly popular for its long prolific flowering. Flowers are dark purple, the undersides of the sepals are marked with a pale silver-grey bar. Young foliage is tinged bronze.
**Flower size:** to 15cm (6in).
**Height:** up to 3m (10ft).
**Blooms:** May to September. Suitable for any position in the garden. Prune lightly in spring.

**VILLE DE LYON**
Large graceful flowers – carmine red edged with dark crimson – on a vigorous plant. Best on a warm wall.
**Flower size:** to 10cm (4in).
**Height:** up to 2.5m (8ft).
**Blooms:** Continuously from July to October. Prefers an east or west facing position. Prune hard in spring.

**VITICELLA PURPUREA PLENA ELEGANS**
This vigorous and free-flowering plant can be trained through tall trees and evergreens, used as ground cover or in containers. The fully double flowers are a dainty mauve colour.
**Flower size:** to 5cm (2in).
**Height:** to 3.5m (11½ft).
**Blooms:** July to September. Suitable for any position. Prune hard in spring.

**VITICELLA RUBRA**
A variety of vi icella good for growing through evergreen and deciduous trees; especially nice with old shrub roses. Flowers are claret red with dark, almost black, anthers.
**Flower size:** under 5cm (2in).
**Height:** 3.5-4m (11½-13ft).
**Blooms:** July to September. Good for any position. Prune hard in spring.

'The President'

'Ville de Lyon'

## VITICELLA ROYAL VELOURS

A versatile, free-flowering plant with wonderful colouring – dark, inky purple with black anthers.
**Flower size:** to 10cm (4in).
**Height:** to 3.5m (11½ft).
**Blooms:** July to September. Good in any position. Prune hard in spring.

## VITICELLA VENOSA VIOLACEA

This variety has deep purple petals, veined with white. All other characteristics as for 'Royal Velours'.

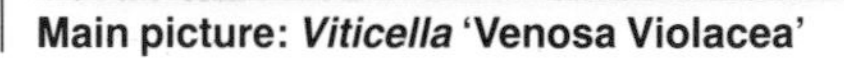

**Main picture:** ***Viticella*** **'Venosa Violacea'** **Inset:** ***Viticella rubra***

**VYVYAN PENNELL**
One of the finest quality double clematis, although the flowers which appear on new growth later in the season are single. They are lavender blue, tinged crimson.
**Flower size:** to 15cm (6in).
**Height:** to 3m (10ft).
**Blooms:** June to August-September. Give a south, east or west facing position. Prune lightly in spring.

**WILLIAM KENNET**
A vigorous grower, covered with huge flowers of a deep lavender blue with a dark reddish bar when first open; stamens are purple. The second crop is equally spectacular. Very free-flowering over long periods.
**Flower size:** to 15cm (6in).
**Height:** to 3.5m (11½ft).
**Blooms:** May and June, with second show in September. Prune lightly in spring.

'Vyvyan Pennell'

## W.E. GLADSTONE

A real curiosity because of its enormous flowers – the largest of any clematis. They are lavender blue with purple stamens.
**Flower size:** can be over 20cm (8in across).
**Height:** to 4m (13ft).
**Blooms:** June to September. Suitable for any position. Prune lightly in spring.

'William Kennet'

'W. E. Gladstone'

# INDEX AND ACKNOWLEDGEMENTS



**Picture Credits**

Gillian Beckett: 4/5, 10(b), 11, 13, 15, 17, 18, 22(l), 32(r), 39(b), 45(t).
David Russell: 28(r), 30(t), 31(l), 32(l), 35(tl), 37(br), 38, 40(bl,tr), 41(b), 42, 43, 44.
Sheila Orme: 47(t).
Harry Smith Horticultural Photographic Collection: 6(b), 10(r), 20, 21, 22(r), 23, 24, 26/27, 28(l), 29, 30(b,r), 31(r), 33(l,br,tr), 34, 35, 36, 37(t), 38(l), 39(t), 41(t), 45(b), 46, 47(b).

**Artwork by**

Richard Prideaux & Steve Sandilands